TYNE & WEAR
archives &
museums

© Tyne & Wear Archives & Museums, 2018

All rights reserved. No part of this book may be reproduced, stored or introduced into a retrieval system or transmitted in any way or by any means (electronic, mechanical or otherwise) withour prior permission of the publishers.

ISBN: 9780905974996

Coal mining using horse power, c.1794. Coal is brought to the surface in baskets called corves, which are emptied into waggons. Waggoners and their waggonway horses are moving the coal down to the river

Contents

Foreword
Introduction
The Evidence
A Difficult and Dangerous Job
The Willington Waggonway
The Archaeology
Conservation
Studying Comparative Waggonways
Analysing the Physical Evidence
The Research Project

Foreword

In the summer of 2013 I was excited to learn that excavations were taking place ahead of the development of part of the old Neptune Shipyard just west of Segedunum Roman Fort in Wallsend. I anticipated that part of the civilian settlement of the fort just might come to light, revealing another piece of the jigsaw of the World Heritage Site at the eastern end of Hadrian's Wall. Little did I imagine that what would be discovered would indeed be a significant piece of world heritage, however, not Roman at all, but instead dating from the late eighteenth century. Even as a self-confessed Romanist I was certainly not disappointed! Indeed, one of my other hats and interests is railways, so unexpectedly finding the earliest example of the international standard gauge was quite simply electrifying, if you'll excuse the railway pun!

The preservation of the section of the wooden Willington Waggonway, including the wash pond, is a remarkable gift to aid our understanding of early railways. The work of the archaeologists, the project team and expert partners deserves great commendation. We must also recognise and thank Arts Council England for its support through the PRISM Fund and the Designation Development Fund that has made this project possible, ensuring the remains are saved for posterity and enabling the research, interpretation and dissemination of this knowledge to an international audience.

This publication, part of that dissemination, presents the incredible story of how the Willington Waggonway was found and saved, and places the remains in their original context within the great network of wooden rail lines restless with waggons moving coal from the mines to ships on the River Tyne. This picture of our early industrial past is now that much clearer thanks to this project. It's also a thought to ponder, especially next time you travel on a train, that the track gauge of just over half of the world's lines conforms to that of the Willington Waggonway, the earliest example to be found, miraculously preserved in urban Tyneside.

Above: The Willington Waggonway excavated in 2013 in Walker, Newcastle upon Tyne with the River Tyne in the background *The Archaeological Practice*

Left: Aerial view of the Willington Waggonway showing the main way on the left and the wash pond with stone paving on the right *The Archaeological Practice*

Introduction

Wooden waggonways were first introduced to the North East of England in the early seventeenth century. Horses pulled waggons full of coal from local pits to staiths at the River Tyne. Waggonways were an extremely efficient form of transport compared to a horse and cart on rough roads, and they allowed coal to get where it was needed faster than ever before. This stimulated an increase in coal production which was at the heart of the economic development of the region. An extensive network of waggonways developed across what was known as the Great Northern Coalfield. These were our earliest railways.

The Willington Waggonway was one of many early wooden railways, so why was its discovery in 2013 so significant?

Not only was it the best preserved and most complete early wooden railway to have been found anywhere in the world, it was also the earliest standard gauge railway to be excavated. Today about 55% of railways in the world are international standard gauge, defined as 4' 8½" or 1435mm.

The excavation at the Neptune Shipyard, in Walker, Newcastle upon Tyne, also unearthed the only 'wash pond' to have ever been professionally excavated and recorded. This would have been used for cleaning and wetting wooden waggon wheels to stop them from splitting or catching fire due to friction. Historians knew that wash ponds existed through documentary sources, but none had been discovered until this excavation.

Re-used ships' timbers were used in the construction or the maintenance of the waggonway. If the types of ship being used could be identified, this could give insight into aspects of maritime history previously unknown to us.

The discovery of the Willington Waggonway was so important because it was identified as an internationally significant artefact, important both for the history of coal mining and as part of the formative history of the railway.

An engraving by Fittler of Newcastle in May 1783 showing coal being transported down a waggonway to the Tyne
Newcastle Libraries

The Evidence

An Archaeologist drawing the stone paving in the wash pond
The Archaeological Practice

Our knowledge of waggonways comes from various sources of evidence. As new information is discovered our understanding of the early history of railways grows.

Books

We are fortunate that waggonways have been a focus of research for many years. In 1970, Dr Michael Lewis published his pioneering book *Early Wooden Railways*. International Early Railways Conferences have been taking place since 1998, each followed up with publication of the proceedings. In 2012, Les Turnbull's *Railways Before George Stephenson: A Study of the Waggonways of the Great Northern Coalfield 1605-1830* affirmed the importance of the North East's wooden waggonways.

Primary Sources

Documents created at the time that events were taking place are primary sources.

A good example is William Gibson's fascinating 'View Book', an engineer's diary from the 1770s. Gibson was part of a team led by coal owner William Brown, known to his contemporaries as 'the father of the coal trade'. Gibson supervised the development of the Willington Colliery, recording his activities in great detail in his View Book which makes it an invaluable primary source.

Entries in 1776 mention the purchase of old ships' timbers for waggonway sleepers. On 17th December 1776, William Gibson was at Sunderland where he 'met Mr Collins about ye old ship wood for Sleepers and marked off 194 pieces and bought some old plank for 25s.'

Evidence of the same trade was found by Amy Flagg who, in her *Notes on The History of Shipbuiding in South Shields 1746-1946* (1979) lists entries in newspapers from the 18th century which advertise the sales of ships' timber.

A page from William Gibson's View Book
The North of England Institute of Mining and Mechanical Engineers

Archaeologists documenting and cleaning the Willington Waggonway
Paul Jarman

Archaeology

Archaeological excavation can tell us a great deal, particularly when matched with descriptions in written sources. The Willington Waggonway excavation has provided crucial evidence that previously was lacking, and the preservation of the timbers gave an opportunity for further study.

Modern Reconstructions

The value of a modern reconstructions such as the one at Beamish Museum should not be underestimated. Made by a wood worker rather than a railway expert, the reconstruction shed light on possible construction methods, it also shows how a horse traverses points and positions itself on curves.

Art

Within the collections of Tyne & Wear Archives & Museums, Northumberland Archives, The Mining Institute and other organisations are artworks depicting waggonways in use. These can be very useful sources of information.

Illustration of Parkmoor Waggonway, Gateshead by Richard Turner
Les Turnbull

Waggon on replica waggonway at Beamish Museum
Paul Jarman

Bringing all the evidence together

All sources must be explored fully and cross-checked to give a clearer picture of the development, operation and maintenance of the waggonway. By doing so we gain a deeper understanding and appreciation of a technology that changed the world.

A Difficult and Dangerous Job

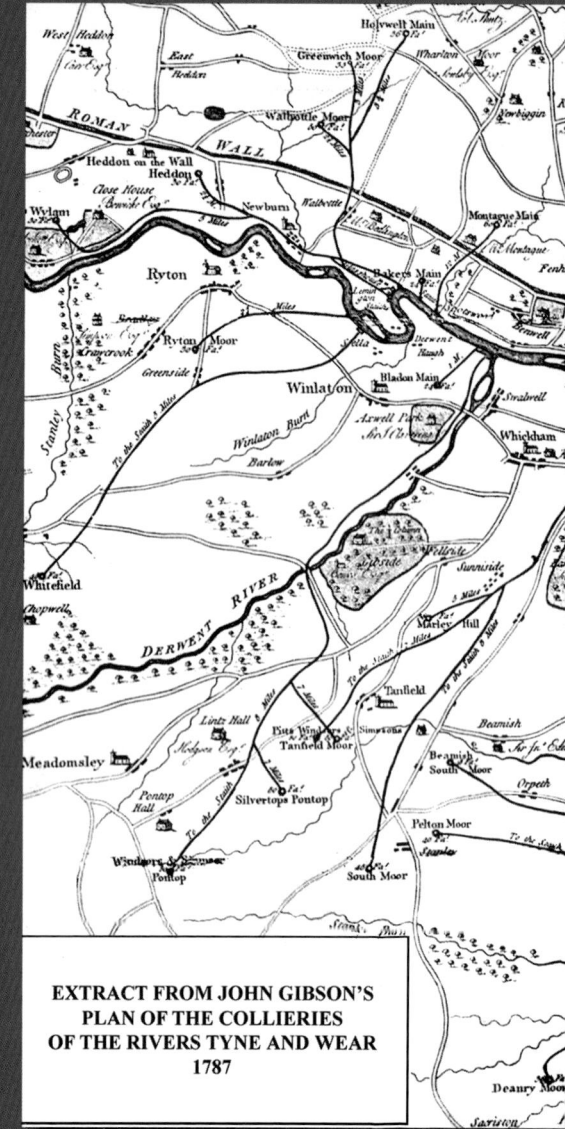

EXTRACT FROM JOHN GIBSON'S
PLAN OF THE COLLIERIES
OF THE RIVERS TYNE AND WEAR
1787

Extract from John Gibson's
Plan of the Great Northern
Coalfield in 1787
Les Turnbull

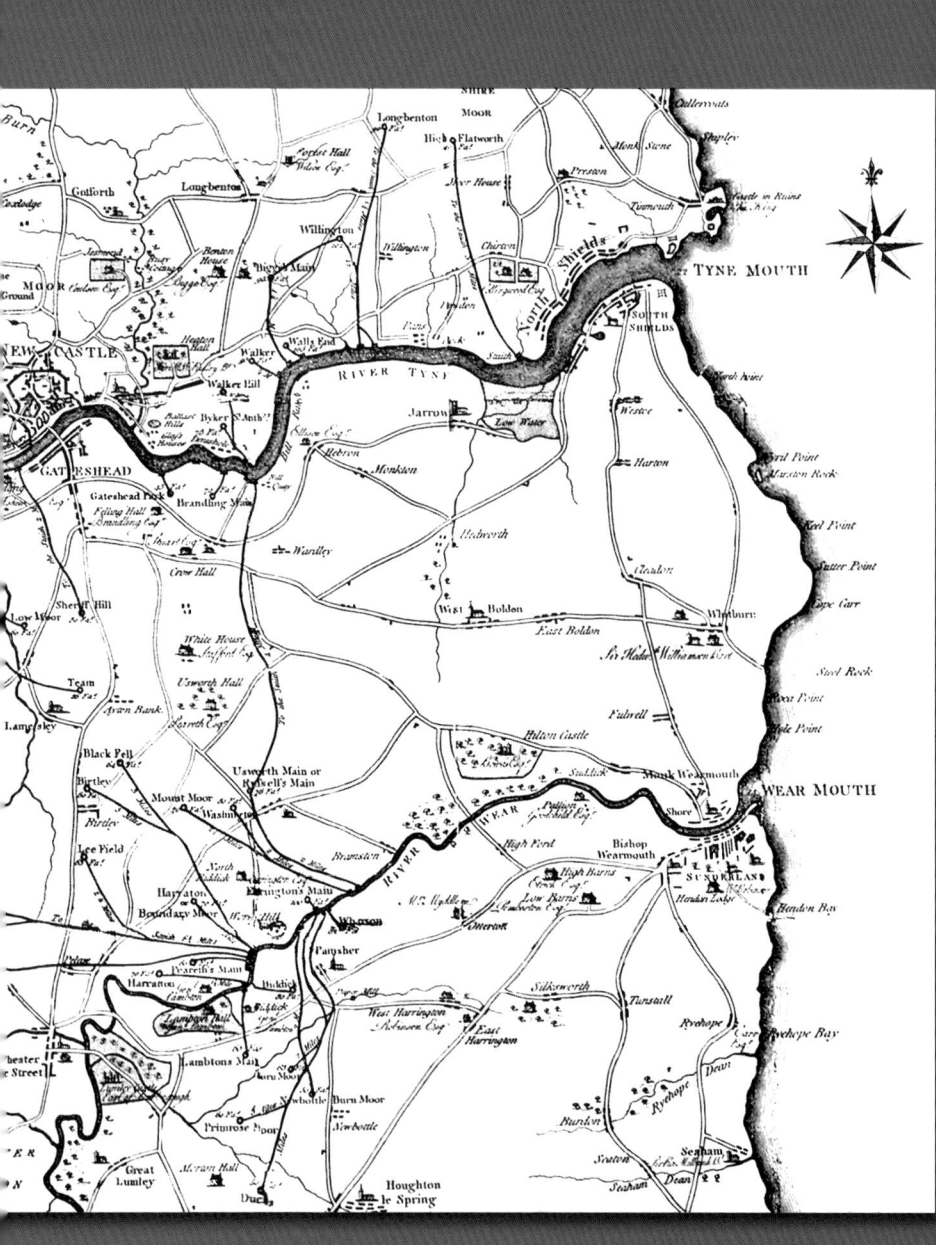

The Coal Trade

For three centuries, from the beginning of the reign of Queen Elizabeth I to the end of the reign of Queen Victoria, the North East of England was the most important coalfield in Britain at a time when the British coal industry was the most important in the world.

Production figures for the coal industry in the eighteenth century show that, for most of that period, the North East of England produced more than a third of the total output from British mines.

Treatise on the Winning and Working of Collieries by Matthias Dunn, 1848, plate 27
Tyne & Wear Archives & Museums

There were three aspects of the coal trade. The landsale trade supplied local needs such as the salt and glass making industries. The export trade mainly supplied the Low Countries and the Baltic. The very important seasale trade primarily supplied domestic fuel to London and the South East of England.

Innovation

The North East was at the forefront of technical innovation and expertise, so the advice of its engineers was in demand all over the world. This led to the development of the railway network in Britain and also in Europe and elsewhere in the world.

Waggonways were the main transport arteries of the largest coal producing area in the world during the eighteenth century. Their construction had many of the complexities of the later iron railways, such as huge earth embankments and spectacular wooden bridges. They represented large investments of capital,

they were built, maintained and operated by a skilled workforce and many were in use for more than a hundred years.

The waggonways were a source of wonder to visitors from elsewhere in Britain and abroad. They even attracted the attention of industrial spies keen to learn the secrets of the 'Newcastle Roads' as the waggonways were referred to by those overseas.

Gosforth Colliery by Thomas Hair
Hatton Gallery, Tyne & Wear Archives & Museums

Willington Colliery

In the area to the east of Willington, the Shire Moor, a coal seam outcrops on the surface and has been mined since medieval times. In Willington, however, the same coal was deep underground. It was beyond the reach of miners until the development of powerful steam pumping engines in the mid-eighteenth century.

Five engines were used to reach the coal at Willington, two large engines for pumping, two smaller engines for returning water which drove the haulage machinery, and a fifth which was built to be portable, known as the Little Engine.

A catalogue of misfortunes noted in William Gibson's diary reveals the unreliable nature of these early steam engines. Building the engines was a difficult task but it was only a prelude to the difficulties encountered in operating these temperamental machines. Gibson and his team also faced many difficulties from quicksand, water, gas and other natural hazards.

Work began at Willington in July 1773 and finally, on 18th October 1775, they "got the coal at Willington Engine Pit". By 1st November 1775 they were close to beginning tunnelling east and west to open up the mine. The next day everyone celebrated, with the owners contributing a fat roasted ox, a large quantity of ale and, incredibly, a waggon of punch.

Low Willington Colliery by Thomas Hair
Hatton Gallery, Tyne & Wear Archives & Museums

Bigges Main Colliery

William Gibson's diaries have given us a lot of information about the winning of Willington Colliery, but very few primary sources have survived which tell us about the operation of it. Bigges Main Colliery, however, was very well documented.

The records provide details of the technical and economic history of the industry and they also give an insight into the living and working conditions of the the mining community of Bigges Main. The story of Willington Square and the colliery would have been similar.

It is clear from documentary evidence that Bigges Main was using the latest technology of the day. Significantly, the cost of the steam engines represented more than a quarter of the capital invested. Although a less significant amount was tied up in the railway, a very large portion was spent on the purchase of horses and their feed. Farms were an important part of the mining enterprise becuase the land provided grazing and food for the horses, which were the main source of power throughout the colliery, including the waggonways. The impact of the Napoleonic Wars on the price of horses and feed at the beginning of the 19th century partly explains why colliery engineers were interested in developing stationary and travelling steam engines to replace horses.

The colliery accounts show that 'binding money' was paid to attract the men to sign an annual bond. This was a contract of employment which stipulated the

A North East Colliery around 1835
Tyne & Wear Archives & Museums

conditions under which the workers were employed. It was often difficult to secure and retain the highly skilled workforce. An allowance was made for the services of a surgeon and for 'subsistence to the sick and maimed pitmen during their illness'. This allowance was common in the northern coalfield and is a reminder of both the dangers inherent in working down the mine and the concern of the owners for the welfare of their workers.

The Workforce

A lot of manpower was required when mining for coal. At Bigges Main Colliery it is estimated that there would have been about 250 men and boys underground and another 150 on the surface. It soon became one of the leading collieries on Tyneside, producing 79,500 tons per year.

Coalmining was a difficult and dangerous job but it was also financially rewarding. Most hewers, for example, would earn much more than agricultural labourers and even some tradesmen. In families where several men and boys were employed the total income could be considerable, allowing them to employ servants.

When the coal industry began to expand in the 19th century, it was the attraction of these high rewards, together with free accommodation and fuel for the household, which motivated many thousands of families from throughout Great Britain and Ireland to move to Northumberland and Durham.

Contemporary opinions of miners weren't always complimentary: *'The people here are little better than savages, and their Countenances bear the marks of hard labour and total ignorance... They eat as well as the substantial tradesmen in great Towns, but they are ragged and dirty, and their wives are idle and drunken so that while they live in plenty they present to your view an air of misery, poverty and oppression...'* – Elizabeth Montagu, owner of East Denton Colliery, letter to Mrs Carter, 31st May 1776.

High wages, high living and uncouth behaviour were, in Elizabeth's mind, the salient features of the mining community.

Dangerous working conditions and high earnings provided an explanation of why so many chose to spend their money and leisure time in escapism; they worked hard and they played hard. But that was only part of the story. There were many forward-thinking and intelligent men from the coal mining communities who rose to positions of prominence, such as George Stephenson who became one of the most distinguished railway engineers of his age.

A Pitman Going to Work by G. Jameson
Tyne & Wear Archives & Museums

The Willington Waggonway

The Coal Waggon
Reproduced with the permission of *Northumberland Archives*

The Route

The Willington Waggonway started as a simple branch line from the Engine Pit which joined the pre-existing Benton Way. It was marked out on August 23rd 1773, showing how important it was to have a railway link at an early stage of development to provide a convenient way of getting materials up to the colliery for the building work.

As the colliery developed westwards, more lines were added until it was a major railway network with an independent route to the Bigges Main staiths. The section of waggonway discovered and excavated in 2013 was part of the Bigges Main line. Although formed from a series of waggonways, the route was collectively known to people in the late 18th and early 19th centuries as The Willington Waggonway.

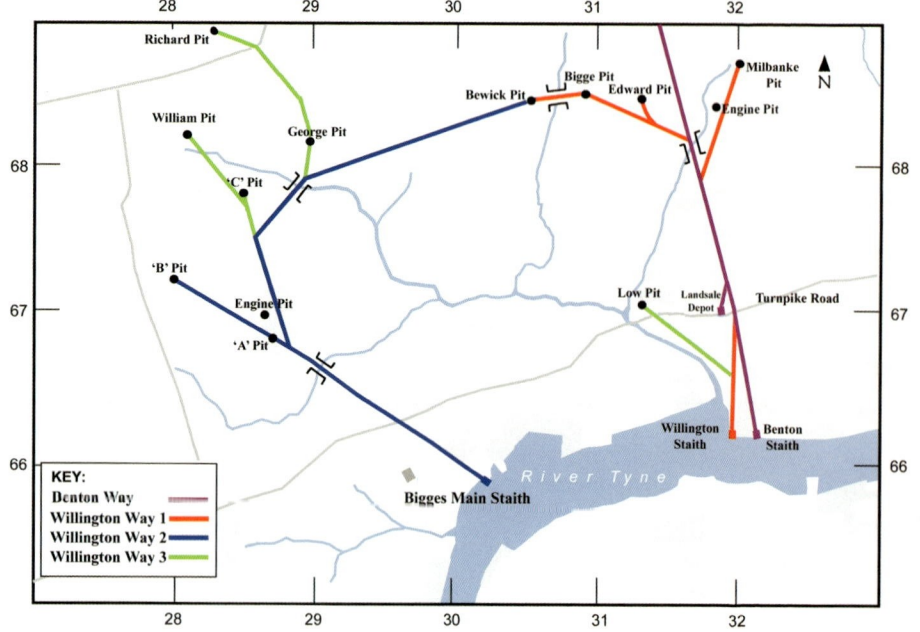

The Development of the Willington Waggonway 1773 – 1810. *Alan Williams*

The later Killingworth Waggonway, which was used by George Stephenson during his development of the steam locomotive, used part of the Willington Waggonway to reach the River Tyne. The gauge of the Willington Waggonway (based on the earlier Benton Way) therefore set the gauge for the Killingworth Waggonway and ultimately most of the world.

Waggons

Coal was measured and sold by volume, not by weight, making it important that the size of the waggon was correct. This measure was known as a chaldron. The size of a chaldron varied in different areas of the coal field. A Newcastle chaldron held fifty-three hundred weight of coal, but we know that Benton waggons were smaller, carring only forty-five hundred weight.

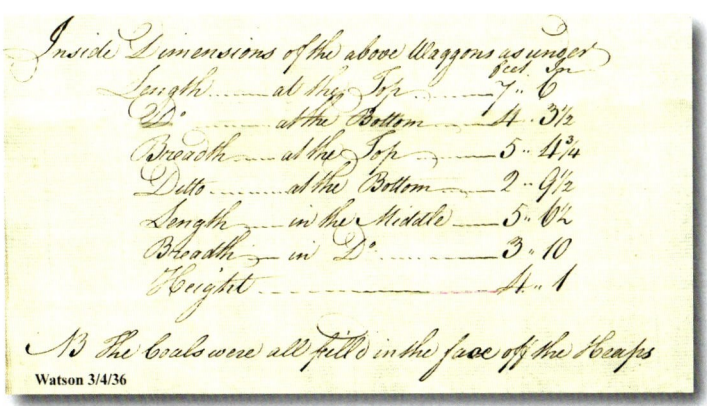

Dimensions of the Willington Waggons
The North of England Institute of Mining and Mechanical Engineers

The waggons initially had four wooden wheels with the only iron work being the axletrees (strong beams which carried a wheel at each end), the fittings and the nails. The wheels and brake suffered the most wear, so a stock of replacements would be held at the colliery. Later developments towards the end of the 18th century led first to iron front wheels and then to all four wheels being made of iron.

A large wooden lever, known as a convoy, was pressed against the wooden wheels by the waggonman to control the waggon on gradients, causing wear to both the convoy and the wheel. Both sometimes caught fire because of friction. Hence there was a need for wash ponds to soak the waggon wheels at the base of steep inclines, such as the descent to Willington Shore. Because of the primitive brakes on waggons at this time, moving heavy loads along a wooden waggonway in wet conditions was a dangerous exercise which caused severe damage to the track.

Usually, the waggons returned empty but sometimes there was an opportunity for the waggonmen to earn additional income with a return load. This could be wood and bricks for the colliery, or manure and lime for farms along the route. The return journey was normally uphill so these would be half loads.

We also have evidence of passengers being carried unofficially on waggonways; there is a record of a domestic servant named Margaret Dobson who was riding on her father's waggon on its descent to Bigges Main village, and was sadly killed when she jumped off. Occasionally there was also official traffic. For example, tourists were encouraged to use the East Kenton Way, Kitty's Drift, for visits to a coalmine.

A Representation of a Coal Waggon
Tyne & Wear Archives & Museums

Horse Power

Horses were only expected to last about five years on the waggonways due to the workload. They required food, shoeing, veterinary services and there was wear and tear on the harness. A horse keeper, who would need rent and free coals, was also required. At Bigges Main Colliery, for example, the annual cost of hay alone was generally three times the maintenance bill for the whole waggonway.

A Representation of a Coal Waggon by John Buddle 1764
Tyne & Wear Archives & Museums

Because the cost of keeping horses was greater than the cost of hiring men, the horses were well looked after. William Gibson's Diary on 17th December 1775 records:

'*being a very bad evening of snow and sleet, I ordered the Engine Pit and also ye Staples to drop work as ye horses cannot stand*'!

Usually, the waggoners were recruited from the neighbouring farms, and often the duty to provide a horse and driver was a condition of a farm's tenancy. Two types of waggoners were employed – those providing their own horses, usually referred to as carriagemen, and those using the company's horses, usually known as drivers.

Waggoners were well paid, as this popular folk song indicates:

> '**Clap hands for daddy coming down the waggonway
> With a pocket full of pennies and a poke full of hay**'.

Recycling and Reuse

Recycling and reuse were very important in the 18th century. In 1771 it was estimated that half the costs of securing the coal at Willington could be raised by reusing existing equipment. There was not the throwaway culture of the 21st century; a lot of expenditure could be saved through the efficient management of existing resources.

Sleepers and rails were often re-used and sometimes large sections of track were lifted and used again elsewhere when a mine became redundant. Old ship timbers were bought in large numbers from breakers' yards for building and repairing waggonways.

Photograph of a re-used ship timber from the Willington Waggonway excavation
The Archaeological Practice

The
Archaeology

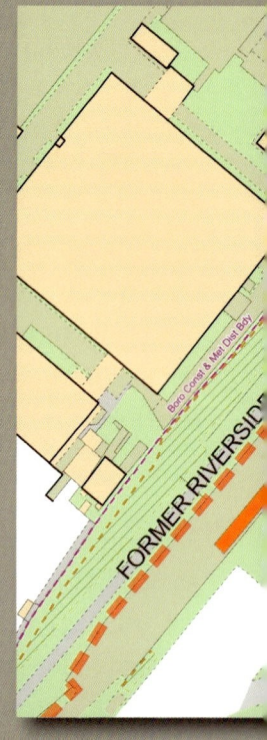

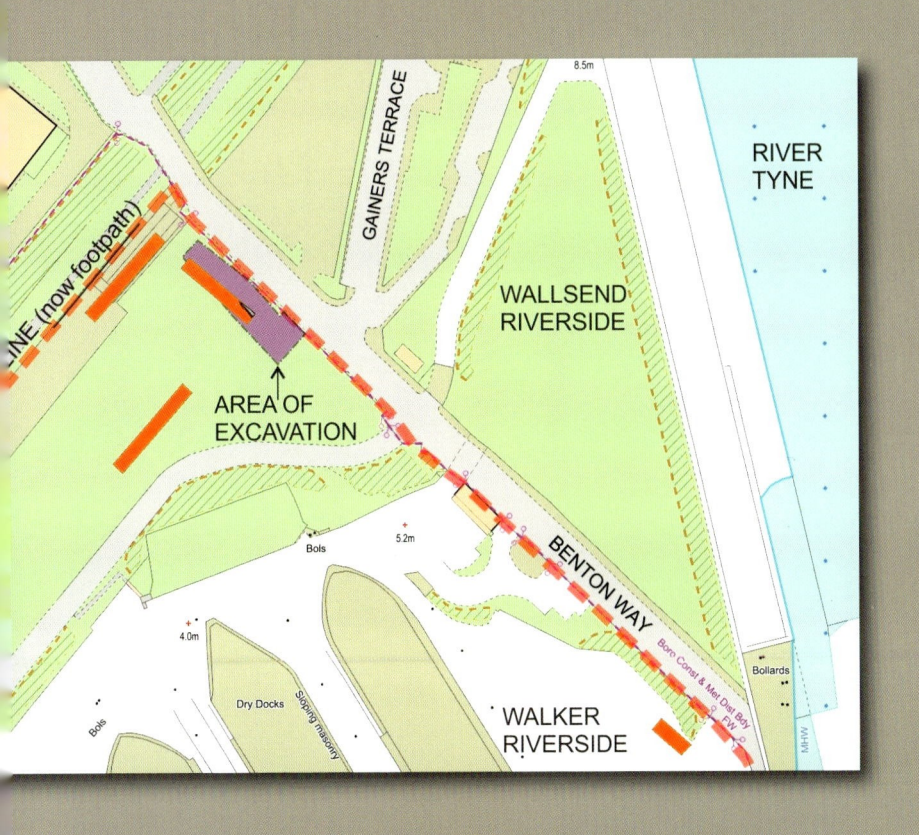

Plan of trenches excavated on the former Neptune Shipyard site. The dotted red line shows the boundary of the site, the red rectangles indicate the trenches excavated in 2013 and the purple rectangle identifies the principal site of excavation

The Archaeological Practice

A Railway Rediscovered

The Willington Waggonway was rediscovered during exploratory archaeological excavations on the site of the former Neptune Shipyard in Walker, Newcastle upon Tyne. These investigations took place due to the close proximity of Segedunum Roman Fort. It was thought that there could be evidence for the Roman civilian settlement or associated burials in the area being developed. The discovery of a wooden waggonway came as a complete surprise.

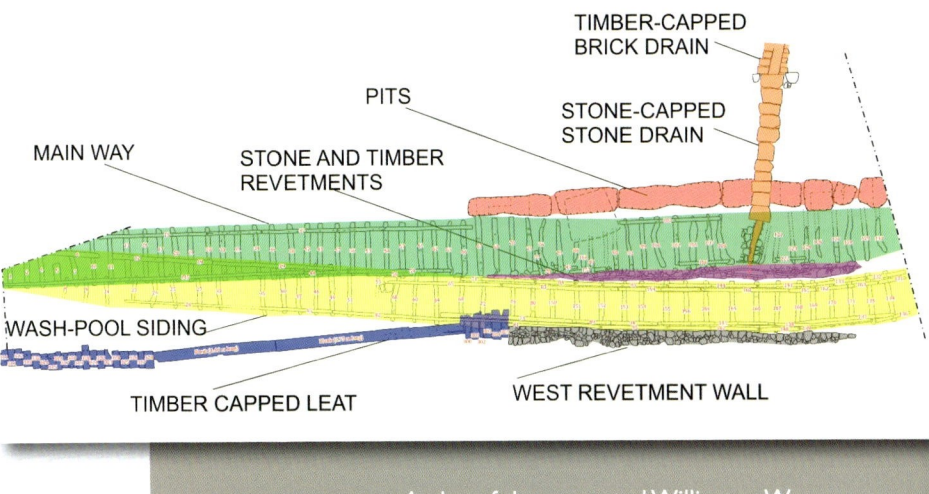

A plan of the excavated Willington Waggonway
showing the key features of the site
The Archaeological Practice

The remains uncovered was a section of wooden railway almost perfectly preserved below a compacted layer of coal waste which had been used as the base of a building previously used for the storage of ships' electric cables. This building occupied the site until the late 20th century. The coal waste which had preserved the waggonway was extremely difficult to remove. The excavation progressed slowly over nearly three months, with the excavators using spades and trowels, occasionally assisted by machine.

The rediscovered Willington Waggonway
The Archaeological Practice

What Was Found

The section of waggonway consisted of a main way along which the coal waggons would have travelled downhill to the staiths, with a separate siding known as a wash pond, constructed in a shallow dip. The wash pond was originally bordered on both sides by stone walls, between which the waggons passed on their way back from the riverside to the mines. The purpose of the lower-lying siding was to contain a pool of water to allow empty waggons to be washed and their wooden wheels soaked, to protect them from drying out and cracking.

The main way consisted of parallel double rails – one wooden rail fixed on top of another – set on wooden sleepers that were embedded in clay and coal waste. All fixings, rail to rail and rail to sleeper, were wooden pegs set in small circular or square holes.

A photograph showing the double rail uncovered during the excavating of the Willington Waggonway
The Archaeological Practice

The double-rail system allowed the upper rail to be replaced easily when it became worn out, without the need to replace sleepers which was a more difficult and costly process. The wash pond was equipped with single rails because empty waggons, travelling more slowly uphill within water, did not cause so much wear to the rails. An interesting feature of the wash pond was a well-made stone trackway within it, held in place by another, narrower set of 'check rails', allowing horses to gain a good foothold when pulling waggons through the water.

A notable feature of the sleepers and rails in the wash pond was that they were the original timbers used to construct the waggonway in or around 1785. The timbers of the main way were in much poorer condition and had probably all been replaced more than once. The majority of the sleepers of the wash pond are of finely cut oak while those of the main way are of various mixed species, some of them parts of unworked tree branches. This shows that there was a need to repair the main way frequently and that timber was an expensive commodity, requiring thrifty solutions. Other features associated with the two sets of tracks were mostly concerned with supplying water to the wash pond and draining it away. Water arrived through a narrow leat, made of stone sides and covered with wooden planks, running down towards the north-bank of the Tyne and connecting obliquely with the high point of the wash pond. Water supplied in this way gathered in the wash pond and drained from it through a controlled exit in the opposite side of the wash pond, at its deepest point. Here, a well-made, stone-capped drain, much larger and more impressive than the leat, drained water directly away from the site in the direction of Benton Way. Originally the entrance to the drain was through a 'revetment', a stone wall built against the embanked main way. Later this was replaced in wood and the remains of a simple sluice mechanism were found in the excavation.

Section drawing showing the construction of the wash pond
Alan Williams

A photograph of the leat with fill removed
The Archaeological Practice

An Enigma

While most of the features of the preserved waggonway are clearly understood in terms of transport, washing of waggons and water supply, one group of features discovered by excavation remains enigmatic. This is a group of six vertically-sided pits, all between 65cm and 99cm wide but varying between 68cm and 3.68m long, arranged in a line alongside the north side of the main way.

These pits seem to have contained wooden stakes in their corners, perhaps for structures to support their walls, and one of them appears to have been cut through by the wash pond drain. There are few other clues to their function. One suggestion is that they held water used to splash on the wheels of waggons, on the side away from the wash pond, making sure that they didn't catch fire due to friction on their way downhill to the staiths. However their purpose remains far from certain.

A photograph showing the mysterious pits
The Archaeological Practice

After the Waggonway

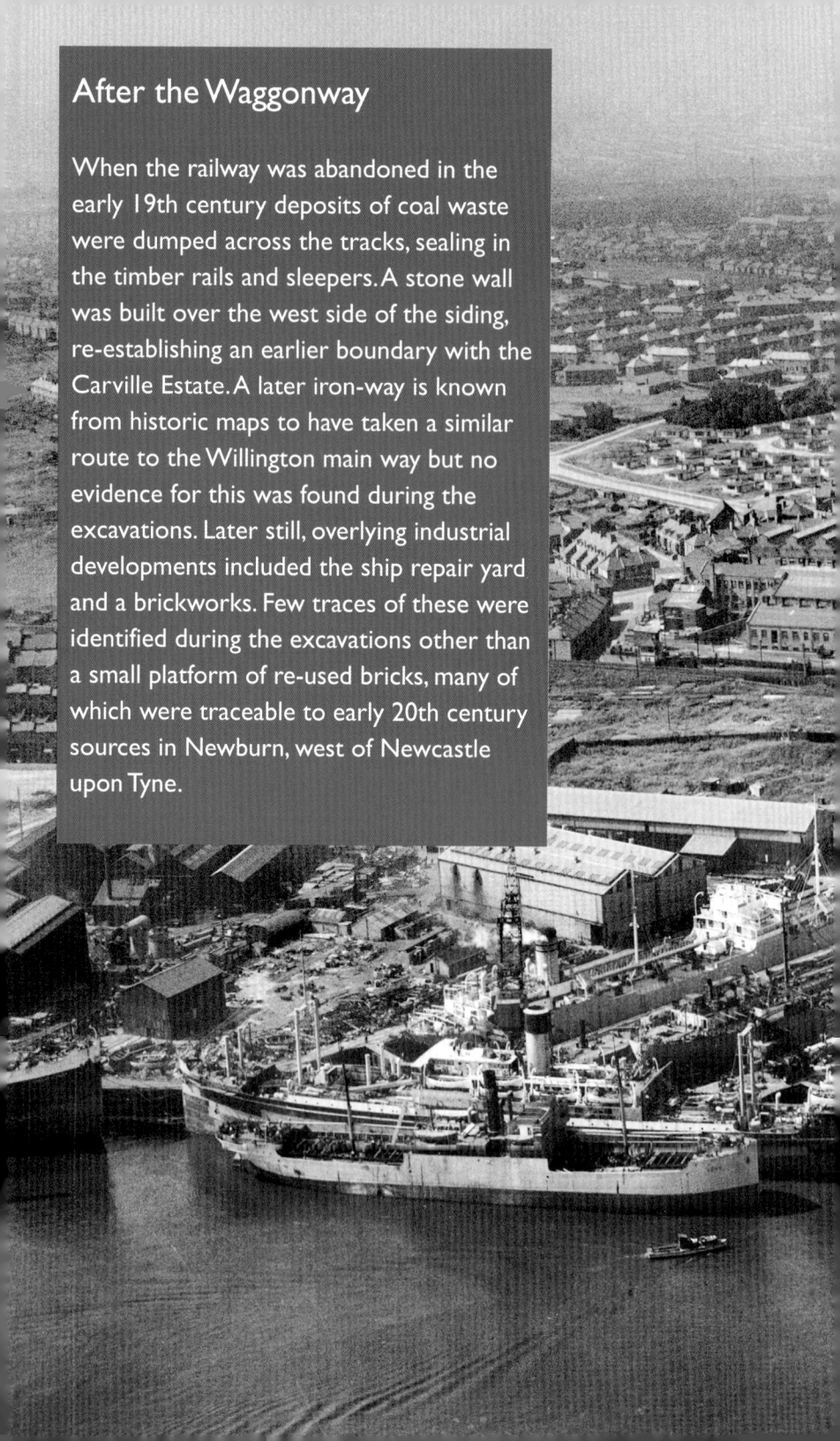

When the railway was abandoned in the early 19th century deposits of coal waste were dumped across the tracks, sealing in the timber rails and sleepers. A stone wall was built over the west side of the siding, re-establishing an earlier boundary with the Carville Estate. A later iron-way is known from historic maps to have taken a similar route to the Willington main way but no evidence for this was found during the excavations. Later still, overlying industrial developments included the ship repair yard and a brickworks. Few traces of these were identified during the excavations other than a small platform of re-used bricks, many of which were traceable to early 20th century sources in Newburn, west of Newcastle upon Tyne.

An aerial photograph of Wallsend, taken on 1st August 1946. The waggonway was underneath the building to the centre right of the photograph, to the left of the road which leads to the ferry terminal

© Historic England

A Miracle of Survival

The discovery of the Willington Waggonway and its recognition as a site of international significance required three 'miracles'. First the survival, hidden and undisturbed, of the remains amid one of the most intensively developed industrial landscapes on Tyneside. Secondly their preservation, especially the wooden components, for more than 200 years. The third miracle was the chance rediscovery of the remains as part of an archaeological intervention in advance of a process of comprehensive reclamation of former industrial land which would undoubtedly have destroyed all of the evidence.

The land all around the concealed waggonway saw massive industrial intervention in the 19th century. Two neighbouring shipyards were established, Swan Hunter's in Wallsend and Wigham

Richardson's in Walker. A brickworks operated for a time. The Riverside Branch Railway arrived immediately to the north to service the expanding industrial activity.

In 1903 the two shipyards combined as Swan Hunter & Wigham Richardson, going on to establish a worldwide reputation as a builder of merchant vessels and warships through most of the 20th century. In 1906 the renowned and record-breaking transatlantic passenger liner RMS *Mauretania*, the largest ship in the world at the time, was launched from a nearby slipway. Thereafter there was an almost continuous procession of naval cruisers, frigates and destroyers, and merchant liners, cable ships and tankers. As shipbuilding and industry grew and declined, the waggonway lay hidden for over two hundred years, waiting to be rediscovered.

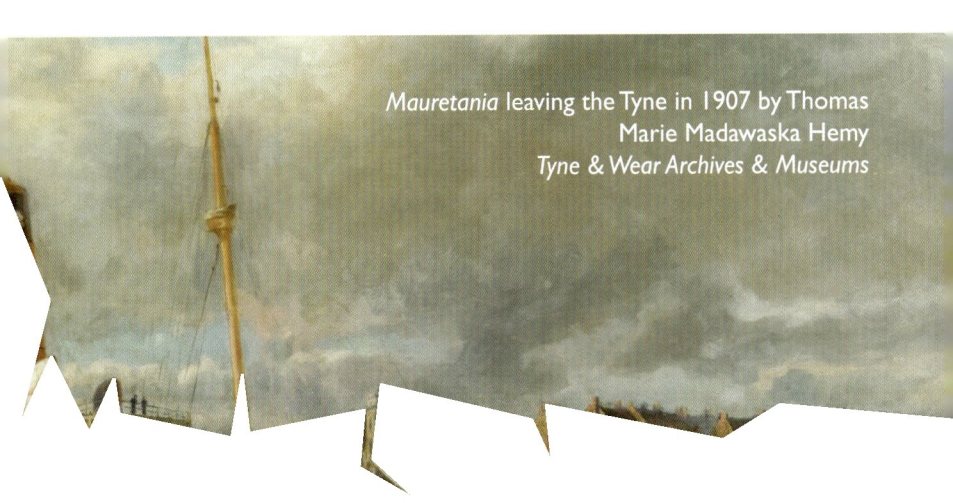

Mauretania leaving the Tyne in 1907 by Thomas Marie Madawaska Hemy
Tyne & Wear Archives & Museums

Significance

The Willington Waggonway was no ordinary line. It comprised a number of features making it the most complete, most complex and best-preserved 18th century waggonway ever excavated. Highly significant and uniquely surviving components of the line are its intact double-rails, previously only known from documentary references and illustrations, and surviving wash pond, the first such feature ever to be excavated. Its association with Killingworth and the origin of the four foot eight and a half inch standard gauge makes it a find of international significance.

Some questions of interpretation remain, such as the date and function of the puzzling pits cut alongside the north side of the main way. However, the broad structural outline and modus operandi of the waggonway are reasonably easy to determine as a result of the unusually good state of preservation and completeness of the remains, and the richness of available documentary material.

An illustration of the main line and wash pond in use
Alan Williams

Conservation

Timber awaiting collection after treatment at York Archaeological Trust

The rescued waggonway timbers
The Archaeological Practice

A Race Against Time

Once the timbers were uncovered we had to act quickly to remove them from the excavation site. Any delays would hold up the redevelopment of the site and exposure to the elements would cause the timbers to begin to deteriorate. Time was of the essence.

Wood from a waterlogged site should be stored in similarly wet conditions until it can be treated. This prevents uncontrolled evaporation of the water from the wood which can lead to the collapse of the cell walls within the structure of the wood, leading to the wood being damaged through shrinking, warping and cracking. This is impossible to reverse; the wood cannot be rehydrated back to its original shape and size.

The waggonway timbers in storage at Stephenson Railway Museum
The Archaeological Practice

Somewhere to store the timbers where uncontrolled air drying could be kept to a minimum needed to be found quickly. The Stephenson Railway Museum, managed by Tyne & Wear Archives & Museums on behalf of North Tyneside Council, has a large area of outdoor space, which was suitable for siting the forty foot storage container required. Metal storage containers are renowned for the build-up of condensation internally, which was ideal as this helped to reduce uncontrolled drying of the timbers.

The staff and volunteers at the museum sprayed the timbers with water several times a week to ensure that they remained wet. However the high levels of condensation also allowed fungal growth to begin to develop on the timbers. When this began to occur, the timbers were sprayed with a 50/50 mixture of Industrial Methylated Spirit and water to kill off the fungal growth and prevent it from spreading.

Waggonway timbers in a tank at York Archaeological trust
Ian Panter

Preserving Waterlogged Wood

A treatment plan was devised for the timbers with advice from York Archaeological Trust (YAT) which specialises in the conservation of waterlogged wood and had provided initial advice prior to the timbers being rescued. During the rescue phase samples were taken and sent to YAT for assessment. These samples indicated that the timbers had survived while buried underground with very little decay. They had not been fully saturated despite being in a waterlogged site and the cell structure had not been badly damaged. Although this initially seemed like good news, it presented a problem. It is actually easier to conserve fully saturated wood, where there has been loss to the cell structure, because the damage allows the chemicals necessary for conservation to more easily penetrate the wood. Wood which has not had a high level of saturation is difficult to conserve as the chemicals cannot fully penetrate.

It was not advisable to allow the timbers to air dry because they would still have been at risk from some degree of shrinkage and warping. The surface of the timbers would have become very fragile and crumbly. YAT recommended that the timbers needed to be treated with Polyethylene Glycol (PEG), a type of wax, followed by controlled drying in a freeze drier.

Treating wood from waterlogged sites using PEG and the process of freeze drying is one of the most effective treatments available. The method was pioneered by the National Museum of Denmark in the late 1960s for the treatment and conservation of the Skuldelev Viking ships. It was also used in Sweden to treat the warship 'Vasa' and more recently has been used to treat and conserve the Tudor warship 'Mary Rose'.

The Treatment Process

We secured enough funding to send 43 timbers from the waggonway to YAT for treatment. When treating wood from waterlogged sites with PEG it is important that the wood is cleaned prior to treatment so that it is free from biological or mineral deposits, such as iron. Failing to remove these deposits can result in the wood having high acidity levels, causing the PEG solution to break down and be much less effective. Once the timbers had been cleaned, they were loaded into the treatment tanks at YAT. Two types of PEG with different molecular weights were added to the tanks in stages. The PEG acts as scaffolding which supports the structure of the wood once the water has been removed.

The timbers were monitored during the time that they were in the tanks to ensure that the PEG had penetrated as far as possible into the timbers. It is important that the PEG fully penetrates the structure of the wood, as PEG not only supports the structure of the wood once it has been dried, it also plays an important part in the freezing process. It prevents the expansion of water contained in the wood as it turns into ice during freezing. If the water is allowed to expand the ice crystals can damage the structure of the wood.

The Final Stages

Once it had been determined that the timbers were saturated as far as possible with both types of PEG they were removed from the tanks and were frozen to -30°C. They were then put into a freeze drier and a vacuum was applied. The temperature in the chamber was slowly allowed to rise, causing the water molecules to turn directly into vapour without becoming a liquid. This vapour was collected in a condenser leaving the wood free from ice or water, resulting in dry timbers. The smaller timbers took approximately two months for the process to be complete and the larger ones took up to four months to become fully dry.

The final stage of the conservation process was to lightly clean the surface of each timber in order to remove any excess PEG which may have congealed on the surface of the timbers. This gives the impression of a white crystal-like substance sitting on the surface which can be unsightly. It is also important to remove any excess PEG from the surface of the wood, because PEG can absorb water, leading to an increase in the moisture content of the newly dried wood. It was important that the newly dried-out timbers were stored in stable environmental conditions, with a stable temperature and relative humidity. Now that the waggonway timbers have been treated and have returned from YAT they are kept in an environmentally controlled store.

Large freeze drier at York Archaeological Trust

Studying Comparative Waggonways

J. Christie Lith. Newcastle.

'A Newcastle Pit in the Olden time'
Newcastle Libraries

The Earliest Wooden Railways

Although not the earliest guided trackways, wooden railways first appeared in the early 16th century in Germany and Austria. The technology was then shared throughout Europe by mining engineers. The first use of a waggonway in Britain appears to have involved carts running on parallel wooden planks steered by a vertical pin on the truck, introduced by German miners at Caldbeck, Cumbria in the 1560s.

The Wollaton Waggonway from Strelley to Wollaton near Nottingham, completed in 1604, is the earliest proven British railway of the modern type (based on the German Riesen system). Others rapidly followed, for example at Broseley & Willey in Shropshire by 1605, and on the Bebside bank of the River Blyth in Northumberland by 1608. On the Tyne, the Whickham Grand Lease Way, opened around 1621, is the first such waggonway evidenced by surviving documents or material remains.

Waggonmen on the Newbottle Way
Sunderland Museum and Winter Gardens

Developments over Time

The type of waggon used on these early waggonways is not known but the design probably developed from the standard coal wains used at the time. Later, the hopper body evolved as the best shape and the North East chaldron measure for coal became a standard measure with waggons built to take that load. The construction of waggonways changed relatively little during the remainder of the 17th century and into the 18th century, when some were built using an L-shaped iron rail system and unflanged wheels, a system known as a plateway. Flat iron plates were also introduced to fix to the top of wooden rails in the later 18th century in order to preserve them. This led to the replacement of wooden wheels with iron in the late 18th century and, subsequently, the replacement of wooden rails with iron, a process hastened from around 1810 by the development of heavy steam locomotives. Wooden waggonways did not entirely die out at this time, however, since many already in existence, as at Lambton 'D' Pit near Sunderland, continued in use for several decades while others were built for local mining operations throughout the 19th century and into the early part of the 20th century.

Archaeological Remains

The early adoption of waggonways in the North East of England, responding to the region's increasing dominance in the British coal mining industry, led to the area's pivotal role in the development of the modern railway, not just in Britain, but worldwide. However, while the history of the railway and the North East's place in it is well documented, there are few records of surviving physical remains derived from archaeological investigations.

Since the mid-1990s, over a dozen excavations of waggonways have taken place in the North East, with others elsewhere, to add to current understanding of how waggonways were built and maintained in practice and how they survive and may be recognised archaeologically.

Studying Comparative Sources

Various research avenues were explored following the excavation and partial conservation of the Willington Waggonway to enhance our understanding of the recorded remains.

A detailed study of the historic background to waggonway construction in the context of North East coal mining was carried out and the results of other waggonway excavations in the valleys of the rivers Tyne and Wear were studied for comparative examples.

About a dozen excavations have taken place on other waggonways in the Tyne and Wear valleys. The earliest remains were on the course of the early 17th century Stella Grand Lease waggonway near Blaydon. The majority of these excavations have produced remains of waggonway embankments and parallel ditches, with impressions of sleepers on top of the embankments but few surviving timbers. By far the best-surviving remains were those of the Lambton 'D' Pit waggonway excavated in 1995-6 at Fencehouses, near Sunderland. At the time of their discovery they were the best preserved and most substantial early wooden railway remains uncovered in the UK.

The Lambton 'D' remains probably date to the period 1812-17 and comprise single-way wooden tracks on oak sleepers laid at 4 feet 3 inch gauge on a bed of ash and coal fragments. The character of rail construction suggested a pragmatic, 'make-do-and-mend' approach involving continual reuse and replacement of timbers. There was evidence of in situ wood-working with wood shavings, discarded dowels and timber off-cuts across the site. An interesting feature of this waggonway was the presence of a crude set of points, without any moving parts, at an intersection between tracks. These are probably comparable to those used on the Willington Waggonway. Also comparable with the Willington Waggonway are the kinds of drain found parallel with and cutting across the waggonway and the scarcity of associated finds such as pottery, clay pipes and horse-shoes.

Researching the Remains of the Willington Waggonway

Detailed studies were carried out on the surviving waggonway timbers, particularly those thought to have been recycled from ship timbers. We know from accounts of wooden waggonway construction that ship timbers were bought for that purpose, so information about possible sources of the timber used on the Willington Waggonway was sought in contemporary adverts for the sale of dismantled ships. In addition, a selection of the timbers was examined after their conservation and compared with contemporary accounts of timber components used in different types of 18th century ship.

The stonework in the horse track within the wash pond was studied in order to discover its likely source of origin. Samples from the unpreserved timbers were also taken for dendrochronology (tree ring analysis) but unfortunately there was not enough information in the samples to provide results.

Willington Waggonway leat capping made of reused ships timbers
The Archaeological Practice

Analysing the Physical Evidence

Excavating the paving of the wash pond of the Willington Waggonway
The Archaeological Practice

An Ideal Source of Timber

The North-East shoreline presents many hazards to ships following the coast, or attempting to enter ports and tidal rivers. Local archives, libraries and museums have handbills and newspaper advertisements for the sale of hulls and materials from ships wrecked or stranded in these waters. An advertisement in the Newcastle Courant of 12th May 1803 provides a direct link between the sale of recycled timbers and the business of waggonway construction:

"To be sold by auction, etc. ...A quantity of timber and oak plank, etc.... suitable for Ship-Builders, House-Carpenters, and for laying Waggon -Ways, &c".

Such auctions of wrecked or stranded ships were advertised regularly in the Newcastle Courant but the trade was to a large degree weather dependent. Easterly gales brought more sales! 1784 was particularly notable for the large numbers of ships wrecked during a great gale from the east that blew from the evening of Sunday 5th December until noon on Wednesday 8th December. Most ships advertised for sale through to January 1785, when the Willington Waggonway is likely to have been under construction, were wrecks from this storm.

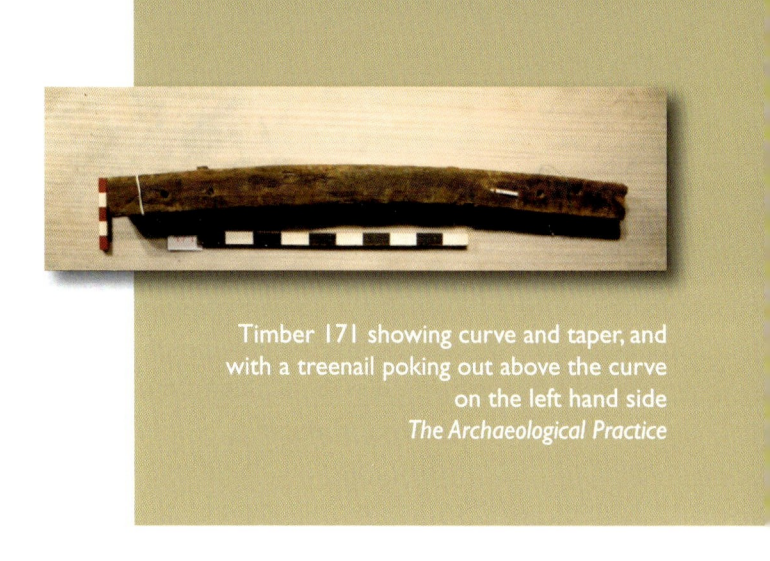

Timber 171 showing curve and taper, and with a treenail poking out above the curve on the left hand side
The Archaeological Practice

The Ships that became The Willington Waggonway

Of the 335 timbers recorded during excavation, 62 were provisionally identified as being originally from ships on the basis of their curved, tapering shape and varying thickness, or evidence of specialised fastenings and joinery. It was impossible to provide more than a tentative identification of original function for the recycled ship timbers. On the basis of their shape two of them may be upper futtocks from the frame of a small vessel such as a sloop, while others appear to be cut from some of the thicker planking found in small to medium sized vessels.

The main indicator of ship origins, however, is the presence of treenails and/or holes bored to take the treenails. Treenails are cylindrical oak pins driven through the planks and timbers of a vessel to connect them together. Once in place they are tightened with one or more wedges to make them waterproof. Treenails were seen in 11 of the 13 conserved timbers available for study.

Left: Illustration of a brig or brigantine from Falconer, 1789 Edition

The treenails found in the Willington timbers suggest regional variation in the method of tightening them, including evidence that at least one was worked on in southern England. However evidence to connect timbers to any particular ship, or shipyard, was not found. In 1785 the North East shipbuilding boom had yet to come to fruition, so the timbers may well not come from a local source. Many vessels operating locally were built in Whitby, London and abroad (including colonial America), although the number of ships being built on the rivers Tyne and Wear was increasing.

Top: Timber 170 treenail head with crossways caulking
Ian Whitehead
Left: Timber 134 treenail tightened with a flat wedge
Ian Whitehead

The Horse Track

Detailed study was made of the stones used to form a compact cobbled surface for a horse-path within the trackway of the wash pond. This provided a firm base for horses pulling waggons through the axle-deep water. Work was carried out to characterise these stones and answer a number of questions: Were the stones sourced locally? Were they quarried or imported specifically for use on the waggonway, or were they a by-product of the mining process which could be serendipitously and cheaply used for this purpose?

All 400 unworked blocks in storage were examined, and 70 of them were cleaned and subject to closer study. The stones range in size from approximately one metre in their longest dimension down to approximately 10cm in their shortest. They are generally uneven in shape. When used to construct the waggonway track, the stones were set with one of their long edges facing upwards. A high percentage of the tops are freshly broken, but some show weathering, where the stone would have been exposed to the elements for many years prior to incorporation in the waggonway. A few of the stones have been very crudely tooled with just a small number of tool marks.

The washed stone subject to close visual examination and technical analysis
Ian Kille

Analysing the Paving

For purposes of analysis, the stones were visually inspected and subjected to various kinds of technical evaluation, including microscopic examination of thin-sections and measurement of surface geochemical characteristics using a portable X-Ray Fluorescence device. These approaches were backed-up with a search for quarries shown on historic maps, other documented information about stone sources, research into the underlying geology of the area, and examination of building materials found in local buildings.

Crated stones from the waggonway in storage *Ian Kille*

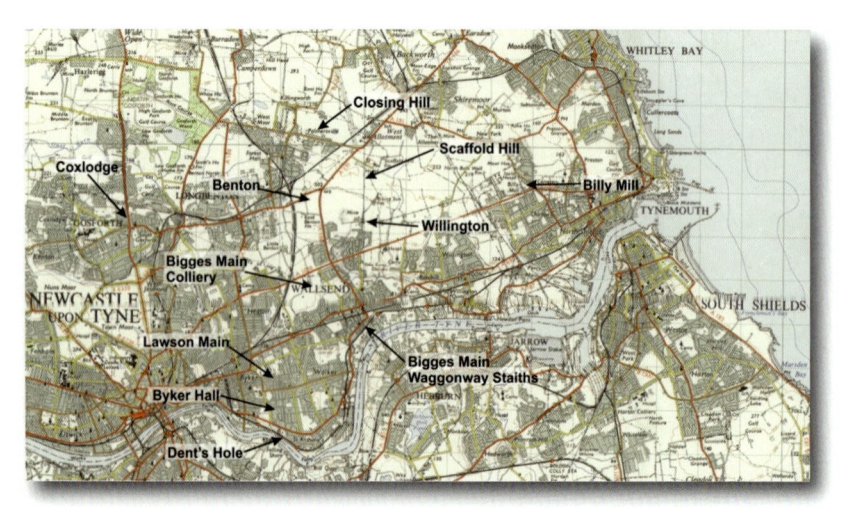

Map of quarries shown on the first edition Ordnance Survey plan (transcribed onto OS 1 inch 7th Series 1955-61)

Example of sandstone showing mica-rich bedding planes, cross-bedding and weathering *Ian Kille*.

The conclusions of this study are that the stones from the waggonway probably come from a single, local source, likely to be the Carboniferous Middle Coal Measures which crop-out in the area and were cut into by coal mines. However, no direct matches for the waggonway stones have been found in local buildings or rock outcrops. A local source is reinforced by documentation in William Gibson's View Book that locally quarried stone was used for other elements of the local mining infrastructure, including the Willington Colliery engine house:

'Work began immediately on building the engine… The stone for this building was quarried locally by the masons.'

Stone from Tynemouth Castle
Ian Kille

Whilst it is possible that the horse-track stones may have been imported by ship, it is highly unlikely they would have been imported for this specific purpose and their importation as ballast also appears unlikely. It is also possible that they were a by-product of coal mining, but several factors make this unlikely. For example, the evidence of weathering on some of the stones which could not, therefore, have come from recently-quarried, underground sources. The proximity of many open quarry sources with properties consistent with the waggonway stones, combined with documentary evidence that locally quarried stone was used to construct mine buildings, strongly suggests that the stones were sourced from a local quarry of which Benton, Willington and Scaffold Hill are the most likely contenders.

Weathering apparent on one of the pieces of sandstone used in the horse-track
Ian Kille

The excavated wash pond with inner check rails removed and stone paving lifted to reveal oak sleeper
The Archaeological Practice

The Research Project

Project Co-ordinator Dominique Bell accessioning the stone paving from the Willington Waggonway into the TWAM collection
Kelly Martin

Previous Work

It is important to note that the initial discovery of the waggonway may not have been followed up so comprehensively were it not for the legacy of three pieces of work.

There have been archaeological excavations in Wallsend since the 19th century exploring the Roman heritage at and around Segedunum Roman Fort. In 1997 the coal mining archaeology of Wallsend was recognised by the discovery, excavation, recording, publishing, preservation and interpretation of the remains of Wallsend 'B' Pit adjacent to the fort.

There was also the excavation, recording and publishing of the extensive remains of a wooden waggonway discovered at the Lambton 'D' Pit, mentioned earlier in this book.

Remains of the Lambton 'D' Waggonway
John Nolan

First held in Durham in 1998, the series of occasional International Early Railways Conferences have stimulated new and renewed lines of research, generating a body of knowledge on the subject. This includes the publication in 2012 of *Railways Before George Stephenson: A Study of the Waggonways of the Great Northern Coalfield 1605-1830* by Les Turnbull.

These excavations and research programmes established the value and practice of researching and recording 18th and early19th century coal mining and waggonway sites in Tyne & Wear. With this record it had become more certain that the opportunity to recover and preserve artefacts for study would be taken, should significant waggonway remains come to light in the future.

Saving the Willington Waggonway

The rescue and preservation of a section of the Willington Waggonway was made possible through the generous and timely support of the Arts Council England PRISM (Preservation of Industrial and Scientific Material) Fund. The PRISM Fund recognised the significance of the discovery and responded quickly, enabling the waggonway to be lifted and stored until further funding could be secured.

Tyne & Wear Archives & Museums collected, by careful and recorded archaeological deconstruction, wooden and stone components within a zone 6 metres in length across the width of the waggonway formation, as well as representative and significant components from other locations on the site.

During the rescue operation it was arranged with colleagues at the National Railway Museum to recover on their behalf a second, smaller group of wooden components from the excavation. The material selected for the national collection based at York consisted of ten timber components from the main way.

The Willington Waggonway Research Programme

Having rescued the waggonway and stabilised the timber components, Tyne & Wear Archives & Museums sought further funding to support a programme of detailed investigation and research into this very significant find. It was recognised that there was much that remained to be understood about the construction, operation and maintenance of early wooden waggonways.

A research programme based on the Willington Waggonway represented a unique and possibily unrepeatable opportunity to bring together the most informed expertise in early railway history and technology to work in partnership on the project.

The Arts Council England Designation Development Fund was considering grants to support projects for 2016-2018 with a theme of 'Research of Designated Collections' making it an ideal time to apply for further funding. The Designation Development Fund has supported the research of the preserved timber and stone components with the project running from December 2016 to March 2018.

The aim of the research project was to understand more about the formative history of the railway through the research and analysis of the excavated remains. This would enhance the documentary evidence to create a new body of knowledge, for publication. The project also engaged visitors through community events, exhibitions, public talks and school visits.

The Excavated Willington Waggonway. *The Archaeological Practice.*

Setting the Standard

This guide book is an introduction to the research undertaken. For those who would like more detail there is *Setting the Standard: research reports on the Willington Waggonway of 1785, the earliest standard gauge railway yet discovered*. This contains all of the research which has taken place to date.

Potential for the Future

The display of the rescued section of waggonway is an ambition for the future. Future capital development at the Stephenson Railway Museum could include the reconstructed remains of the Willington Waggonway as a key display within a redeveloped museum. Potentially there could also be a reconstruction of the waggonway outside, complete with a working wash pond. This would complement the original remains and the models within the museum.

In terms of interpretation at the original site of the waggonway, there are plans for information boards and signage indicating the actual location. There is also perhaps an opportunity to reference the waggonway, and its significance for railway lines around the world, at the nearby Wallsend Metro station.

Currently there are approximately fifty timbers which have been rescued but not preserved. These are subject of further funding requests and a public appeal. By purchasing this guide book you have supported this appeal and the legacy of the Willington Waggonway.

The Willington Waggonway Research Programme Logo (Illustration by Alan Williams)
Tyne & Wear Archives & Museums

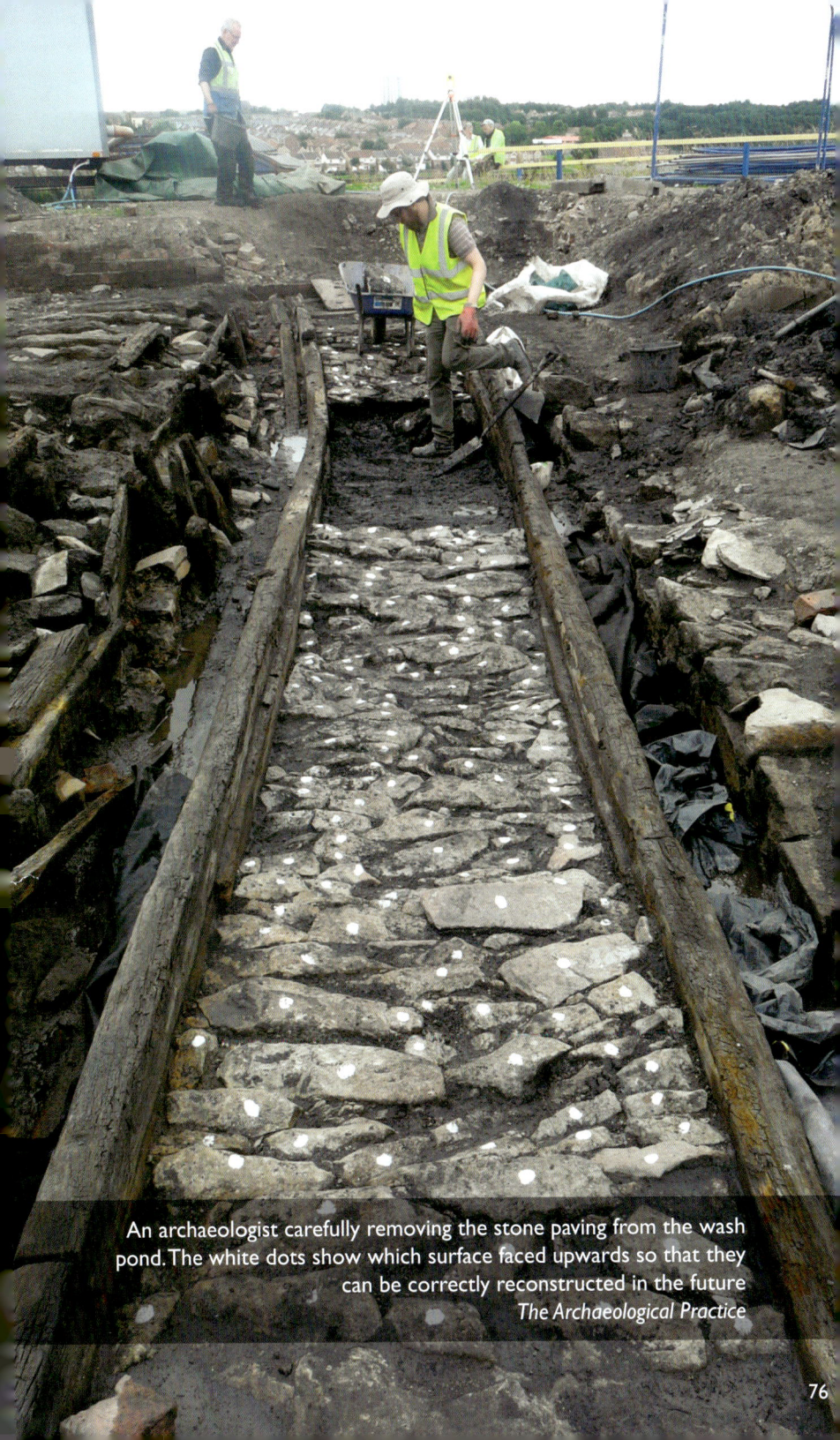

An archaeologist carefully removing the stone paving from the wash pond. The white dots show which surface faced upwards so that they can be correctly reconstructed in the future
The Archaeological Practice

Acknowledgements

We would like to thank Arts Council England for their generous support.

The waggonway timber and stone components were rescued and preserved through a grant from the PRISM (Preservation of Industrial and Scientific Material) Fund.

The Willington Waggonway Research Programme was funded by The Designation Development Fund.

We would also like to thank everyone who offered support and advice during the project.

Contributing authors:
Les Turnbull, Richard Carlton, Alan Williams, Dominique Bell, John Clayson, Rachael Metcalfe, Geoff Woodward, Ian Kille, Ian Whitehead

Published by:
Tyne & Wear Archives & Museums

Publication editor:
Dominique Bell